21st Century
Basic Skills
Library

TALLY CHARTS

Ball Colors		Total
Blue	|||	3
Red	≠≠	5
Yellow	||	2
Pink	||||	4

by Sherra G. Edgar

Cherry Lake Publishing • Ann Arbor, Michigan

2

Published in the United States of America by Cherry Lake Publishing
Ann Arbor, Michigan
www.cherrylakepublishing.com

Consultants: Janice Bradley, PhD, Mathematically Connected
Communities, New Mexico State University; Marla Conn, Read-Ability
Editorial direction and book production: Red Line Editorial

Photo Credits: Eugene Sergeev/Shutterstock Images, cover, 1, 4; Brand X
Images/Thinkstock, 6; Fara Spence/Shutterstock Images, 8; iStockphoto/
Thinkstock, 10, 12, 14, 16; Fuse/Thinkstock, 20

Library of Congress Cataloging-in-Publication Data
Edgar, Sherra G.
  Tally charts / Sherra G. Edgar.
      pages cm. -- (Let's make graphs)
  Audience: K to grade 3.
  Includes bibliographical references.
  ISBN 978-1-62431-395-0 (hardcover) -- ISBN 978-1-62431-471-1
(paperback) -- ISBN 978-1-62431-433-9 (pdf) -- ISBN 978-1-62431-509-1
(ebook)
  1. Graphic methods--Juvenile literature. 2. Tallies--Juvenile literature.
3. Charts, diagrams, etc.--Juvenile literature. I. Title.

QA90.E358 2013
518'.23--dc23

                                          2013005089

Cherry Lake Publishing would like to acknowledge the work of The
Partnership for 21st Century Skills. Please visit *www.p21.org* for more
information.

Printed in the United States of America
Corporate Graphics Inc.
July 2013
CLFA11

TABLE OF CONTENTS

What Is a Tally Chart?

Emma wants to count her money. She can use a **tally chart**. Tally charts help people count.

My Money

		Total
Dimes	////	3
Quarters	//////	6
Dollars	////	3

Tally charts keep track of **data**. Emma can use the data in a **graph** later.

Dog Colors

		Total
Green	~~||||~~ | |	7
Orange	| | | |	3
Blue	| | | | |	4

The **tally** looks like a line.
It stands for an **amount**.
One tally stands for one of
something. We write tallies in
groups of five.

Making a Tally Chart

Nick likes milk. What drinks do his friends like? He made a tally chart to find out.

My Friends' Drinks

Milk	
Juice	
Water	

Nick drew the tally chart.
He wrote in three drinks.

My Friends' Drinks

Milk	////
Juice	//////
Water	//

Nick asked ten friends what drinks they like. He put a tally next to each drink.

My Friends' Drinks

		Total
Milk	////	3
Juice	/////	5
Water	//	2

Nick counted the tally marks. Then he wrote the **totals**.

My Friends' Drinks

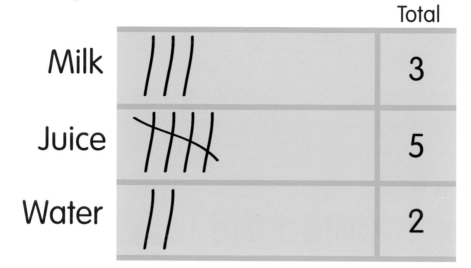

		Total
Milk	///	3
Juice	////	5
Water	//	2

My Friends' Drinks

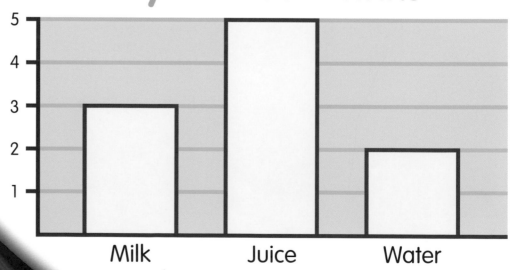

Here is Nick's tally chart. Nick can make a graph with the data. Nick's friends like juice best.

You Try It!

Ask your friends what sports they like. Make a tally chart of the sports. Can you make a graph with the data?

Find Out More

BOOK

Murphy, Stuart J. *Tally O'Malley*. New York: HarperCollins, 2004

WEB SITE

Brain Pop Jr. Tally Charts and Bar Graphs
http://www.brainpopjr.com/math/data/tallychartsandbargraphs/
Watch a video to learn about tally charts and graphs.

Glossary

amount (uh-MOUNT) how many or how much there is of something

data (DEY-tah) facts from a graph

graph (GRAF) a picture that compares two or more amounts

tally (TAL-ee) a mark that stands for an amount

tally chart (TAL-ee CHAHRT) a table that uses marks to keep track of data

total (TOHT-l) the whole amount

Home and School Connection

Use this list of words from the book to help your child become a better reader. Word games and writing activities can help beginning readers reinforce literacy skills.

a	for	make	they
amount	friends	marks	three
an	graph	milk	to
ask	groups	money	totals
asked	he	next	track
best	help	Nick	use
can	her	of	wants
chart	here	one	we
charts	his	out	what
count	in	people	with
counted	is	put	write
data	it	she	wrote
do	juice	something	you
drew	keep	sports	your
drink	later	stands	
drinks	like	tallies	
each	likes	tally	
Emma	line	ten	
find	looks	the	
five	made	then	

Index

About the Author

Sherra G. Edgar is a former primary school teacher who now writes books for children. She also writes a blog for women. She lives in Texas with her husband and son. She loves reading, writing, and spending time with friends and family.